2024 TOYOTA RAV4 HYBRID CAR REVIEW

Your In-Depth Guide to the Car's Interior Features, Pricing, Driving Experience, Safety Features, Trim Levels, Performance and Ownership Costs

JAMES AUTOTRENDS

Table of Contents

Introduction

Overview of the 2024 Toyota RAV4 Hybrid

The 2024 Toyota RAV4 Hybrid stands as a testament to the brand's commitment to innovation, combining a versatile and popular SUV model with a cutting-edge hybrid powertrain. As the automotive industry continues to transition towards more sustainable and fuel-efficient options, the RAV4 Hybrid emerges as a frontrunner, offering a compelling package that seamlessly integrates eco-friendly technology with the practicality and style synonymous with the RAV4 nameplate. The hybrid variant, in particular, showcases Toyota's dedication to providing consumers with greener alternatives without compromising on performance or utility.

Building on the success of its predecessors, the 2024 RAV4 Hybrid inherits the renowned reliability and robust design that has made the RAV4 a staple in the SUV market. The hybrid model brings forth a 2. 5-liter naturally aspirated petrol engine paired with an electric motor, delivering a commendable system output of 160kW. This sophisticated hybrid system not only enhances fuel efficiency but also contributes to a more environmentally conscious driving experience. The RAV4 Hybrid has become a symbol of Toyota's forward-thinking approach, aligning with global efforts to reduce carbon footprints and promote sustainability.

Brief Mention of Market Trends and Changes Since the Previous Generation

In the automotive landscape, the period since the launch of the previous generation RAV4 in 2019 has witnessed significant shifts in market dynamics and consumer preferences. The most notable trend is the accelerating adoption of hybrid and electric vehicles, driven by an increased awareness of environmental issues and a growing desire for eco-friendly transportation. The 2024 RAV4 Hybrid enters this landscape as a response to these changing dynamics, aligning with the market's demand for cleaner and more sustainable driving solutions.

Moreover, the market has experienced fluctuations in pricing, influenced by various economic factors and supply chain challenges. The RAV4 Hybrid, once launched at a starting price of $41, 140, reflects the broader trend of rising vehicle prices across the industry. Despite the challenges posed by external forces, Toyota remains dedicated to delivering a compelling value proposition with the RAV4 Hybrid, emphasizing the integration of advanced technology, safety features, and a comfortable driving experience.

As consumers navigate an evolving automotive market, the 2024 RAV4 Hybrid stands as a symbol of adaptability, embodying the brand's commitment to staying at the forefront of industry trends while meeting the diverse needs of modern drivers.

Chapter 1: Pricing and Value

Pricing Evolution from Launch in 2019 to Current Prices

The pricing evolution of the 2024 Toyota RAV4 Hybrid provides a fascinating snapshot of the dynamic forces shaping the automotive market over the past few years. Launched in 2019 with a starting price of $41, 140 plus on-road costs, the RAV4 Cruiser 2WD Hybrid has undergone a notable transformation in terms of its financial standing. Today, in 2024, the price has escalated to $51, 410 before on-road costs, illustrating the impact of various economic factors and market trends.

This upward price trajectory is reflective of the broader industry trend of increasing vehicle costs, influenced by factors such as inflation, supply chain disruptions, and advancements in technology. Despite the price hike, the RAV4 Hybrid maintains its appeal by offering a plethora of features and capabilities, making it a competitive option in its segment. The evolution in pricing also underscores Toyota's commitment to delivering a high-quality, hybrid SUV that caters to the evolving needs of consumers.

Comparison with Rivals in the Market

In a market teeming with diverse SUV offerings, the 2024 Toyota RAV4 Hybrid faces competition from various rivals, each vying for consumer attention. One notable contender is the GWM Haval H6 Ultra Hybrid, presenting a compelling option at $45, 990 drive-away. The H6 Ultra Hybrid not only challenges the RAV4 Hybrid in terms of pricing but also offers a commendable driving experience, making it a strong alternative for budget-conscious consumers.

Another competitor worth considering is the Nissan X-Trail e-Power, available in Ti spec for $54, 690. While differing in hybrid technology, the X-Trail e-Power provides a unique proposition with its petrol engine acting solely as a generator, showcasing the diversity in hybrid approaches within the SUV market.

As the RAV4 Hybrid's price has increased considerably, potential buyers might find themselves contemplating these alternatives. The pricing landscape becomes crucial in the decision-making process, with consumers weighing the value proposition of each model against their individual preferences and requirements. Toyota's pricing strategy, while reflective of industry trends, necessitates a careful consideration of the overall package and features to maintain a competitive edge in a market where discerning consumers have an array of choices.

Chapter 2: Interior Features and Comfort

Overview of the Interior Design

The interior design of the 2024 Toyota RAV4 Hybrid encapsulates a harmonious blend of smart functionality, ergonomic design, and a commitment to passenger comfort. Toyota has consistently excelled in crafting interiors that cater to the needs of diverse drivers, and the RAV4 Hybrid's cabin is no exception. The design ethos revolves around simplicity, creating a welcoming and intuitive space for both the driver and passengers.

A hallmark of the RAV4's interior design is the thoughtful layout, providing easy access to controls and a well-organized dashboard. Quality fit and finishes permeate the cabin, underscoring Toyota's commitment to delivering a premium experience. From the choice of materials to the strategic placement of features, the interior design reflects a meticulous attention to detail, contributing to an overall enjoyable driving and riding experience.

Upgrades and Changes in the Most Recent Update

The most recent update to the RAV4 Hybrid brings forth several noteworthy upgrades, enhancing the interior's modernity and functionality. A standout feature is the introduction of a larger 10. 5-inch touchscreen media system, a considerable leap from the 8. 0-inch screen available in previous iterations. This not only adds a touch of contemporary flair but also improves visibility and accessibility for higher-spec versions.

The digital instrument cluster has also undergone a transformation, now measuring 12. 3 inches. This upgrade provides a crisp, colorful, and more modern interface, elevating the overall driving experience. Despite its advanced capabilities, the interface remains user-friendly, with intuitive controls accessible through the steering wheel.

Details on the New Touchscreen and Digital Instrument Cluster

The new 10. 5-inch touchscreen in the 2024 RAV4 Hybrid represents a significant leap forward in infotainment technology. With a higher resolution and brighter display, it adds a touch of sophistication to the interior. However, the absence of physical buttons on the edges of the screen and a slightly more complex menu system may require a brief acclimatization period. Despite these nuances, the touchscreen integrates seamlessly with features like Apple CarPlay and the JBL sound system, providing an immersive in-cabin experience.

The 12. 3-inch digital instrument cluster is a technological focal point, offering a modern and customizable interface. Navigating through menus using the steering wheel controls is intuitive, and the information displayed is not only useful but also adds a futuristic touch to the dashboard. These technological upgrades affirm Toyota's commitment to staying at the forefront of in-car entertainment and information systems.

Comfort and Adjustment of Seats

The 2024 RAV4 Hybrid prioritizes passenger comfort, particularly evident in the Cruiser trim. The driver's seat, equipped with 10-way power adjustment including lumbar support, ensures an ergonomic and customizable driving position. While the passenger seat provides eight-way power adjustment, it may pose a challenge for taller individuals due to its elevated position. However, the inclusion of heated and ventilated seats, conveniently controlled near the gear selector, adds an extra layer of comfort.

Despite potential height constraints, the rear seat accommodations shine in terms of space and versatility. Ample headroom, generous knee room, and comfortable width allow adults to sit comfortably, and the inclusion of ISOFIX points, top-tether points, and rear air vents enhance the rear-passenger experience. The smaller sunroof, while appreciated, may seem modest compared to contemporary standards.

Storage Options and Convenience Features

The RAV4 Hybrid impresses with its thoughtful storage solutions, ensuring a clutter-free and organized interior. The large shelf in front of the passenger provides ample space for personal items, and strategically placed nooks near the gear selector and large cup holders enhance convenience. Additionally, the covered center console bin, spacious door pockets with bottle holders, and a flip-down armrest in the rear add to the overall practicality.

Notably, the interior caters to modern connectivity needs with three front and two rear USB ports. The inclusion of Apple CarPlay (wired and wireless) and Android Auto (wired) further enhances the convenience of in-cabin entertainment and connectivity. While the new touchscreen may have some complexities, the inclusion of physical controls for essential functions, such as climate control, ensures that crucial features remain easily accessible and user-friendly. The overall interior design and convenience features affirm the RAV4 Hybrid's commitment to providing a comfortable and technologically advanced driving environment for both the driver and passengers.

Chapter 3: Performance and Powertrain

Detailed Information on the Hybrid Powertrain

The heart of the 2024 Toyota RAV4 Hybrid lies in its advanced hybrid powertrain, a sophisticated integration of a 2. 5-liter naturally aspirated petrol engine, an electric motor, and a small battery pack. This combination delivers a commendable system output of 160kW, providing a powerful yet efficient driving experience. The hybrid system operates seamlessly through an e-CVT (continuously variable transmission), showcasing Toyota's expertise in hybrid technology.

One of the key strengths of the hybrid powertrain is its versatility in different driving scenarios. The system intelligently switches between electric-only mode, petrol engine operation, or a combination of both, optimizing fuel efficiency and minimizing environmental impact. The

RAV4 Hybrid excels in urban environments, often utilizing electric power during coasting and standstill moments. The transition between power sources is smooth, contributing to a refined and enjoyable driving experience.

Comparison with the 2. 0L Four-Cylinder Petrol Engine

For those seeking an alternative to the hybrid powertrain, the 2024 RAV4 offers a 2. 0-liter four-cylinder petrol engine. This engine, with a power output of 127kW and 203Nm of torque, provides a robust urban driving experience. While not as environmentally conscious as the hybrid counterpart, the 2. 0L engine excels in real-world efficiency, making it an attractive option for drivers primarily navigating city streets.

However, it's crucial to note that the 2. 0L four-cylinder petrol engine may not be the ideal choice for towing enthusiasts. The maximum towing capacity for 2WD Hybrid models, equipped with the hybrid powertrain, is limited to 480kg (including braked towing). In contrast, other RAV4 models, including the 2. 0L and AWD Hybrid variants, offer higher towing capacities ranging from 750kg (unbraked towing) to 1500kg (braked towing). Therefore, consumers with towing requirements should carefully consider their needs and opt for the engine configuration that aligns with their usage.

Towing Capacities and Considerations

For those considering the 2024 RAV4 Hybrid for towing purposes, it's essential to understand the model-specific towing capacities and considerations. The 2WD Hybrid models, while excelling in fuel efficiency and urban performance, have a maximum towing capacity of 480kg, making them suitable for lighter towing needs. On the other hand, the AWD Hybrid models, with an additional 'eFour' electric motor at the rear axle, boast higher towing capacities, offering flexibility for a broader range of towing scenarios.

Unbraked towing capacities vary between 750kg for all other models and 800kg for 2. 0L models. The AWD Hybrid and 2. 5L petrol models stand out with a robust braked towing capacity of 1500kg, providing ample capability for towing trailers, small boats, or other recreational gear.

Consumers should align their choice of RAV4 model with their specific towing requirements, keeping in mind the intended usage and the weight of the load. Whether opting for the hybrid efficiency or the conventional petrol power, the RAV4 offers a towing solution for various preferences and needs.

Chapter 4: Driving Experience

Impressions of the RAV4's Driving Manners

The driving manners of the 2024 Toyota RAV4 Hybrid elevate it to a position of prominence in the competitive SUV market. With sweet and natural feeling steering, the RAV4 makes maneuvering through city streets and navigating tight parking spaces a breeze. The steering is well-weighted, offering a sense of control and precision, whether cruising at pace on the highway or executing parking maneuvers.

On lumpy roads and uneven surfaces, the suspension of the RAV4 demonstrates admirable prowess. It adeptly absorbs shocks and handles sharp edges, contributing to a smooth and comfortable ride. The driving experience is characterized by predictability and responsiveness, making it an ideal choice for families seeking a versatile and enjoyable SUV for daily commuting and weekend adventures.

Hybrid Powertrain Performance and Efficiency

The hybrid powertrain of the 2024 RAV4 Hybrid is a standout feature, providing a compelling blend of performance and efficiency. The system's 2. 5-liter naturally aspirated petrol engine, electric motor, and small battery pack work in harmony, delivering a system output of 160kW. The hybrid powertrain showcases its intelligence by seamlessly transitioning between electric and petrol modes, optimizing fuel efficiency in various driving scenarios.

In urban environments, the RAV4 often operates in electric-only mode during coasting and utilizes battery power for smooth acceleration from standstill. This intelligent power management contributes not only to impressive fuel efficiency but also to a lower environmental impact. The RAV4 Hybrid stands as a testament to Toyota's commitment to providing a powerful yet eco-friendly driving experience, meeting the demands of environmentally-conscious consumers.

Noise Levels and Overall Driving Comfort

While the RAV4 Hybrid excels in various aspects, it is not without its nuances. Noise levels in the cabin, particularly during engine kick-ins under hard throttle, can be noticeable. Additionally, some ambient noise from the tires may intrude at certain speeds and surfaces. However, the most prominent noise source tends to be wind rush around the mirrors and windshield at higher speeds.

Despite these considerations, the overall driving comfort of the RAV4 remains commendable. The cabin is designed to minimize noise intrusion, offering a serene environment for occupants. The suspension plays a crucial role in smoothing out road imperfections, ensuring a comfortable ride for both the driver and passengers. As a family-oriented SUV, the RAV4 excels in providing a refined and enjoyable driving experience, balancing performance with comfort.

Safety Features and Driving Assistance Technology

The 2024 RAV4 Hybrid is equipped with a comprehensive suite of safety features and driving assistance technology, reaffirming Toyota's commitment to passenger safety. The active safety technology includes a subtle lane-keeping assistance system for highway driving, offering a gentle steering input to keep the vehicle centered within the lane. Stop-start cruise control further enhances the driving experience by alleviating the strain of low-speed traffic.

The RAV4 Hybrid boasts an array of safety features such as autonomous emergency braking, pedestrian/cyclist detection, blind-spot monitoring, lane departure warning, lane tracing assist, and rear cross-traffic alert. These features contribute to the RAV4's impressive five-star ANCAP safety rating, earned in 2019 and valid until December 2025. The safety-centric design, combined with advanced driving assistance technology, positions the

RAV4 as a reliable and secure choice for families and individuals prioritizing safety in their driving experience.

Chapter 5: Trim Levels and Features

Overview of the Cruiser Trim

The Cruiser trim of the 2024 Toyota RAV4 Hybrid stands as the epitome of luxury and sophistication within the RAV4 lineup. Positioned as a high-spec variant, the Cruiser trim combines advanced technology, premium comfort, and a host of convenience features to offer an elevated driving experience. The Cruiser trim caters to discerning drivers who seek not only the efficiency of a hybrid powertrain but also a premium and well-appointed interior.

Highlighted Features in the Cruiser Trim

The Cruiser trim of the 2024 RAV4 Hybrid boasts an array of features that accentuate its premium status. Notable among these features are the 18-inch alloy wheels for the Hybrid variant, enhancing both aesthetic appeal and on-road performance. The inclusion of a power sunroof adds a touch of luxury, providing an open and airy feel to the cabin. Heated and ventilated front seats, coupled with leather upholstery, elevate the comfort level, ensuring a pleasant driving experience in various weather conditions.

Technological advancements are prevalent in the Cruiser trim, with a 10. 5-inch infotainment touchscreen that serves as the centerpiece of the dashboard. This high-resolution display is complemented by a 12. 3-inch digital instrument cluster, providing a modern and customizable interface for the driver. The JBL sound system, ambient lighting, and a surround-view camera contribute to the immersive in-cabin experience, making every journey in the Cruiser trim a luxurious affair.

Comparison with Other Trims in the RAV4 Lineup

As the flagship trim, the Cruiser stands out with its extensive feature set, but it's crucial to compare it with other trims in the RAV4 lineup to provide a comprehensive understanding of available options. The RAV4 lineup includes various trims such as GX, GXL, and XSE, each catering to different preferences and requirements.

Compared to the entry-level GX trim, the Cruiser offers a significant upgrade in terms of features, comfort, and technology. While the GX serves as a practical and budget-friendly choice, the Cruiser goes beyond with premium materials, enhanced connectivity, and additional safety features.

In comparison to the GXL trim, the Cruiser maintains its luxury appeal but introduces more advanced features such as the power sunroof and a larger infotainment screen.

The GXL strikes a balance between affordability and a well-equipped interior.

The XSE trim, on the other hand, focuses on a sportier aesthetic and driving experience. While it shares some technological features with the Cruiser, the XSE prioritizes a more dynamic and performance-oriented driving experience.

Ultimately, the choice between trims depends on the driver's preferences, budget considerations, and the desired level of luxury and technology. The Cruiser, as the top-tier trim, offers a compelling package for those seeking the pinnacle of features and sophistication within the RAV4 lineup.

Chapter 6: Safety and Reliability

ANCAP Safety Rating and Breakdown

The 2024 Toyota RAV4 Hybrid has undergone rigorous testing by the Australasian New Car Assessment Program (ANCAP) and achieved the highest possible safety rating of five stars. This rating, valid until December 2025, attests to the RAV4's commitment to passenger safety across various categories.

In the ANCAP breakdown, the RAV4 demonstrated excellence in adult occupant protection with a score of 93%, showcasing the effectiveness of its structural design and safety features. Child occupant protection received a commendable score of 89%, emphasizing the RAV4's suitability for family transportation. Vulnerable road user protection and safety assist categories scored 85% and 83%, respectively, reflecting the RAV4's dedication to not only the well-being of its occupants but also pedestrians and other road users. The comprehensive ANCAP safety rating underscores the RAV4's reliability and commitment to delivering a secure driving experience.

Standard Safety Features Across the Range

The 2024 RAV4 lineup is equipped with an extensive array of standard safety features, ensuring that safety is prioritized across all trims. These features contribute to the RAV4's impressive ANCAP safety rating and make it a reliable choice for safety-conscious consumers.

Key standard safety features include autonomous emergency braking (AEB), pedestrian/daytime cyclist detection, intersection turn assist, adaptive cruise control, blind-spot monitoring, lane departure warning, lane tracing assist, rear cross-traffic alert, reversing camera, front and rear parking sensors, traffic sign recognition, trailer sway control, emergency steering assist, and hill-start assist control. The inclusion of seven airbags, including front, front side, full-length curtain, and a driver's knee airbag, further enhances the comprehensive safety profile of the RAV4.

Specific Safety Features in the Cruiser Trim

As the flagship trim, the Cruiser elevates the safety features to a higher standard, reinforcing its position as a top-tier and secure choice. In addition to the standard safety features present across the RAV4 lineup, the Cruiser trim introduces advanced safety technologies that further enhance the driving experience.

The Cruiser trim is equipped with a 360-degree surround-view camera, providing a comprehensive view of the vehicle's surroundings. This feature aids in parking and maneuvering in tight spaces, offering an additional layer of convenience and safety. The inclusion of a digital rear-view mirror enhances visibility by displaying a clear, unobstructed view of the rear, especially useful when cargo or passengers obstruct the traditional mirror's view.

While the standard safety features create a robust safety foundation, the Cruiser trim's additional technologies contribute to an enhanced sense of confidence and security on the road. The safety-centric design and reliability of the RAV4, especially in the Cruiser trim, make it a standout choice for those prioritizing safety in their SUV purchase.

Chapter 7: Ownership Costs

Warranty Details

Toyota provides owners of the 2024 RAV4 Hybrid with a comprehensive warranty package, ensuring peace of mind and long-term reliability. The standard warranty includes coverage for five years or an unlimited number of kilometers, showcasing Toyota's confidence in the durability and quality of their vehicles. This warranty provides reassurance to owners, covering unforeseen issues and instilling confidence in the overall ownership experience.

In addition to the standard warranty, Toyota offers an extended powertrain cover for up to seven years, contingent upon adhering to logbook servicing conducted by authorized mechanics. This extended coverage reflects Toyota's commitment to supporting owners and underscores their confidence in the longevity of the RAV4 Hybrid's powertrain components.

Servicing and Maintenance Costs

The servicing and maintenance costs of the 2024 RAV4 Hybrid are designed to be transparent and manageable, contributing to the overall affordability of ownership. The service intervals are set at 12 months or 15, 000 kilometers, allowing owners flexibility in scheduling maintenance appointments. Toyota offers a capped-price servicing plan for the first five years or 75, 000 kilometers, with an average maintenance cost during this period amounting to a reasonable $260.

Adhering to Toyota's Hybrid powertrain protocols not only ensures regular maintenance but also extends the powertrain cover to up to seven years. The inclusion of an annual health check for petrol-electric models provides additional assurance, and the warranty coverage for the hybrid battery pack extends up to 10 years when following the recommended servicing protocols.

Fuel Consumption and Real-World Efficiency

The 2024 RAV4 Hybrid excels in fuel efficiency, making it an economical choice for environmentally-conscious drivers and those seeking to minimize fuel expenses. The claimed fuel consumption figures for the RAV4 Hybrid are impressive, with the 2WD Hybrid model achieving 4. 7L/100km and the AWD Hybrid model slightly higher at 4. 8L/100km.

Real-world efficiency, as experienced during driving tests, aligns closely with these claimed figures. Over a week of driving, the RAV4 Hybrid demonstrated a commendable 5. 7 liters per 100 kilometers, showcasing its ability to deliver efficient performance in diverse driving conditions. The hybrid powertrain's intelligence in seamlessly transitioning between electric and petrol modes contributes to this notable fuel efficiency, making the RAV4 Hybrid a practical and cost-effective choice for daily commuting and longer journeys.

Owners can appreciate not only the fuel savings but also the reduced environmental impact of driving a hybrid vehicle. The RAV4 Hybrid aligns with the global shift toward sustainability, offering a compelling combination of fuel efficiency and responsible environmental stewardship. The low fuel consumption and real-world efficiency of the RAV4 Hybrid contribute to its appeal as an economical and eco-friendly SUV choice.

Conclusion

Summarization of Key Points Discussed

In the exploration of the 2024 Toyota RAV4 Hybrid, several key aspects have been highlighted to provide a comprehensive understanding of this SUV. The RAV4 Hybrid, with its hybrid powertrain, emerges as a standout choice for those seeking a balance of performance, efficiency, and environmental responsibility. From the evolution of pricing since its launch in 2019 to the detailed examination of the interior features, performance capabilities, and safety aspects, each facet contributes to the RAV4 Hybrid's appeal.

Pricing has evolved considerably since the RAV4 Hybrid's debut, with a notable increase in current prices. Despite this, the Cruiser trim, as the pinnacle of the lineup, introduces a host of premium features, emphasizing its position as a top-tier choice. The interior design, characterized by smart layouts, quality finishes, and upgraded technology, enhances the overall driving experience. The inclusion of a larger touchscreen and digital instrument cluster in the most recent update reflects Toyota's commitment to staying at the forefront of automotive technology.

The hybrid powertrain, a key highlight, offers a powerful yet efficient driving experience, seamlessly transitioning between electric and petrol modes. The RAV4 Hybrid's driving manners, comfort levels, and safety features contribute to its reputation as a family-friendly SUV. The ANCAP safety rating of five stars, coupled with a breakdown of specific safety features, underlines the RAV4's commitment to passenger safety.

Ownership costs are addressed through a comprehensive warranty, transparent servicing costs, and impressive fuel efficiency. The capped-price servicing plan and extended powertrain cover contribute to the overall affordability and long-term reliability of the RAV4 Hybrid.

Final Thoughts on the 2024 Toyota RAV4 Hybrid

The 2024 Toyota RAV4 Hybrid emerges as a compelling choice in the competitive SUV market. It successfully combines the efficiency of hybrid technology with the practicality and versatility expected from a family-oriented SUV. The Cruiser trim, with its premium features, showcases the RAV4's potential to cater to drivers seeking both sophistication and functionality.

As a daily driver, the RAV4 Hybrid excels in providing a comfortable and enjoyable experience, with a strong emphasis on safety and advanced technology. Its hybrid powertrain not only contributes to reduced fuel consumption but also aligns with the growing global emphasis on eco-friendly driving solutions.

In conclusion, the 2024 Toyota RAV4 Hybrid stands as a well-rounded SUV, offering a harmonious blend of performance, efficiency, safety, and comfort. While pricing may have evolved over the years, the RAV4 Hybrid continues to be a strong contender in its segment, appealing to drivers who prioritize a modern, environmentally conscious, and family-friendly driving experience.